ESSENTIAL PHYSICS STAGE 9

FOR CAMBRIDGE SECONDARY 1

WORKBOOK

Kevin Lancaster

Viv Newman | Editor: Lawrie Ryan

Name: ....

Class: ....

School: ....

OXFORD

UNIVERSITY PRESS

OXFORD
UNIVERSITY PRESS

Great Clarendon Street, Oxford, OX2 6DP, United Kingdom

Oxford University Press is a department of the University of Oxford.
It furthers the University's objective of excellence in research, scholarship,
and education by publishing worldwide. Oxford is a registered trade mark of
Oxford University Press in the UK and in certain other countries

Text © Kevin Lancaster, Viv Newman and Lawrie Ryan 2013
Original illustrations © Oxford University Press 2015

The moral rights of the authors have been asserted

First published by Nelson Thornes Ltd in 2013
This edition published by Oxford University Press in 2015

All rights reserved. No part of this publication may be reproduced,
stored in a retrieval system, or transmitted, in any form or by any
means, without the prior permission in writing of Oxford University
Press, or as expressly permitted by law, by licence or under terms
agreed with the appropriate reprographics rights organization.
Enquiries concerning reproduction outside the scope of the above
should be sent to the Rights Department, Oxford University Press, at
the address above.

You must not circulate this work in any other form and you must
impose this same condition on any acquirer

British Library Cataloguing in Publication Data
Data available

978-1-4085-2077-2

10 9

Printed in Great Britain by Ashford Colour Press Ltd., Gosport

Acknowledgements

Cover photograph: © Getty Images/Eric Meola
Illustrations: Tech-Set Ltd
Page make-up: Tech-Set Ltd, Gateshead

Although we have made every effort to trace and contact all
copyright holders before publication this has not been possible in all
cases. If notified, the publisher will rectify any errors or omissions at
the earliest opportunity.

Links to third party websites are provided by Oxford in good faith
and for information only. Oxford disclaims any responsibility for
the materials contained in any third party website referenced in
this work.

Contents

Introduction   iv

Chapter 1   Electricity   1

Exercise 1	Static electricity – dangers and uses	1
Exercise 2	Simple electrical circuits	3
Exercise 3	Circuit symbols and electrical measurements	4
Exercise 4	Modelling current and potential difference	6
Exercise 5	Mains electricity	9

Chapter 2   Moments, pressure and density   10

Exercise 1	Forces and simple machines	10
Exercise 2	Levers and moments	12
Exercise 3	Moments and equilibrium	14
Exercise 4	Pressure on surfaces	16
Exercise 5	Pressure in liquids and gases	17
Exercise 6	Density of fluids and hydraulics	18

Chapter 3   Thermal energy transfers and the energy crisis   20

Exercise 1	Thermal energy	20
Exercise 2	Conduction and convection	21
Exercise 3	Radiation	23
Exercise 4	Evaporation	25
Exercise 5	Fuels for electricity	26
Exercise 6	Renewable energy resources	28

Welcome to Science for Cambridge Secondary 1!
Stage 9 Physics Workbook

Using this workbook

This 'write-on' workbook contains fun, interesting and challenging exercises to help you to develop the skills you need to do well in science. It is divided into exercises that support the topics in the Stage 9 Physics student book, and is arranged in the same order. You will find that looking back at the student book helps you to complete the exercises.

Your teacher can set these exercises as homework or as in-class activities. Your teacher may ask you to work alone, or in pairs or groups. For some exercises, your teacher may ask you to report back to your class on what you found out.

There are simple instructions at the start of each exercise, and a clear description of what the exercise will help you learn. There is space for you to write your answers and for your teacher to mark your work and give you some advice and feedback. Your workbook becomes a record of your progress.

Teacher feedback

Your teacher will be able to find suggested answers to the exercises on their Teacher's CD.

Chapter 1 Electricity

Exercise 1: Static electricity – dangers and uses
Student Book Topics 1.1, 1.2 and 1.3

This exercise is all about static electricity. You will be learning how objects become charged and what effect this can have.

1 Complete the following sentences.

   a When two different materials are rubbed together they become _____ .

      This is called _____ electricity.

   b There are two types of charge called _____ (+) and _____ (−).

   c Charges of the same type _____ each other.

   d Charges of different types _____ each other.

   e When two different materials are rubbed together, _____ are rubbed off one material and on to the other.

   f Each electron carries a small _____ (−) electrical charge.

   g The material that gains electrons becomes _____ charged.

   h The material that loses electrons becomes _____ charged.

2 When a balloon is rubbed onto clothes made from wool, it becomes negatively charged.

   a Explain why it becomes negatively charged.

      _____

      _____

   b Why does the wool end up with a positive charge?

      _____

      _____

   c The negative charge on the balloon is exactly equal to the positive charge on the jumper. Why is this?

      _____

      _____

   d When the balloon is brought close to a wall, it can stick to the wall even though the wall is not charged. Complete the diagram above to show what happens to the negative charges in the wall to make the balloon stick.

   e What clue from the diagram tells us that the balloon is negatively charged?

      _____

1 Electricity

3 The following sentences explain how a photocopier uses static electricity to copy pages of text and images. The sentences are in the wrong order. Number the sentences to explain the way in which a photocopier works. The first one has been done for you.

	The paper is heated to make the black powder stick.
	An image of the page to be copied is projected on to the plate.
	The black powder is transferred from the plate to a sheet of paper.
	There is now a copy of the original page.
	The parts of the plate that are still charged attract bits of black powder.
1	A copying plate is electronically charged.
	Where light falls on the plate, the electrical charges leak away.

4 When driving down a road, charge can build up on a petrol tanker. Explain what causes the charge to build up.

_____

_____

5 Petrol tankers have a large pipe through which the fuel is transferred. Explain the way in which static charge can build up as petrol flows through the pipe.

_____

_____

6 When static charge builds up, there is a risk of a spark. Explain what can be done to prevent sparks.

_____

_____

7 Can you think of another situation where static electricity can be a hazard?

_____

8 Static electricity is also used when cars are painted.

Explain the advantage of painting a car using static electricity. Your answer should include the terms in the box.

| positive   negative   attract   even coating |
| reducing waste   charge |

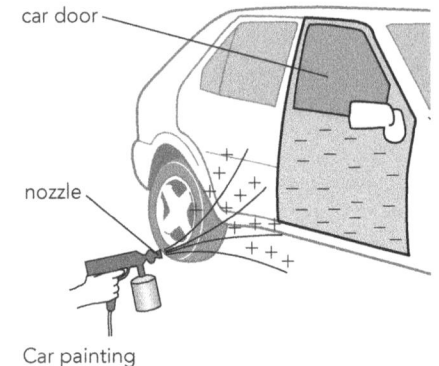

Car painting

_____

_____

Exercise 2

Simple electrical circuits
Student Book Topic 1.4

This exercise will help you to find out what you know about simple electrical circuits.

1 Complete the following paragraph using the words below. Each word can only be used once.

| move | electrons | current | electricity | conductors |

A conductor allows _____ to pass through it. In solid conductors, an electric _____ is a flow of _____ . Metals are good _____ of electricity because some of the electrons from their atoms can _____ freely throughout the metal structure.

2 Have a look at the objects around you. Write the names of five objects that will conduct electricity and five objects that will not conduct electricity in the table below.

Will conduct electricity	Will not conduct electricity

3 To produce a current, a potential difference needs to be applied across a conductor. Potential difference is measured in volts. What provides the potential difference in portable electronic devices?

_____

4 One of the simplest electric circuits can be found in a torch.

Torch

Add the three labels in the box to the diagram.

| switch | lamp | cell |

5 Some metals are better conductors than others. One of the best is gold, but copper is more commonly used in electrical wiring. Why is copper used instead of gold?

_____

Teacher feedback

_____

3

Exercise 3

Circuit symbols and electrical measurements

Student Book Topics 1.5 and 1.6

In this exercise you will be practising drawing circuit diagrams, as well as finding out what you know about series and parallel circuits and measuring electricity.

1 Draw circuit diagrams for the following circuits in the spaces provided.

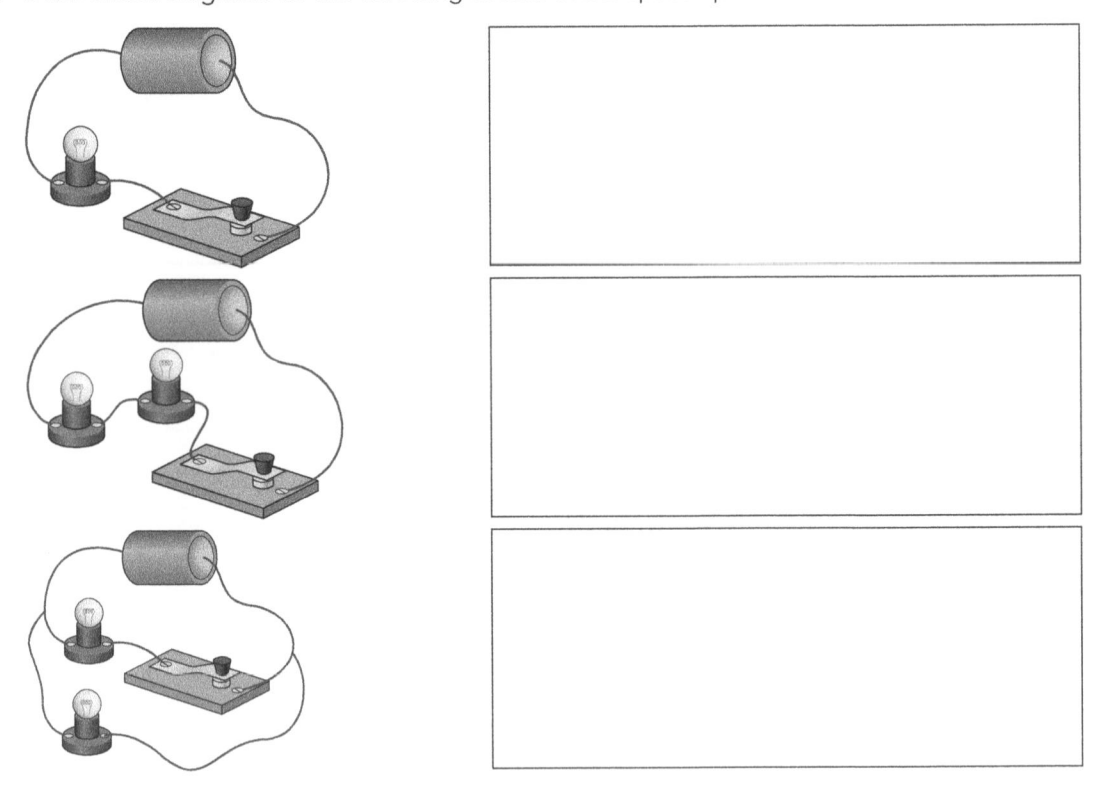

2 Complete the following sentences. One describes a series circuit, the other a parallel circuit.

   a A _____ circuit is one in which the current does not have a choice of path.

   b A _____ circuit is one in which the current has a choice of path.

3 The table below contains statements about series and parallel circuits. Complete the table by writing either **series** or **parallel** next to each statement.

Statement	Series or parallel?
The current is the same in each component.	
If one lamp fails, the others stay on.	
Lights in a house are wired in this type of circuit.	
The components are connected to the battery by one loop of wire.	
Current splits and there are different routes back to the battery.	
If one lamp fails, the others go out.	
Adding more lamps doesn't affect the brightness of the lamps.	
Adding more lamps will make the lamps dimmer.	

1 Electricity

4 For each of the following, circle the correct answer.

  a Current is measured in: **degrees / celsius / amperes / joules / volts**.

  b Potential difference is measured in: **degrees / celsius / amperes / joules / volts**.

  c Potential difference is measured using: **an ammeter / a cell / a thermometer / a voltmeter**.

  d Current is measured using: **an ammeter / a cell / a thermometer / a voltmeter**.

  e Voltmeters are connected: **in series / in between the cells / in parallel / outside the circuit**.

  f Ammeters are connected: **in series / in between the cells / in parallel / outside the circuit**.

5 Look at the first circuit diagram you drew on the previous page. Redraw the diagram, adding the devices used to measure current and potential difference.

6 All bulbs have resistance. The bigger the resistance, the harder it is for a current to flow through a circuit. Explain what happens to the current in a series circuit as more bulbs are added.

_____

_____

Teacher feedback

Exercise 4

Modelling current and potential difference

Student Book Topics 1.7 and 1.8

This exercise will help you to gain an understanding of what electricity is. It will also help you to understand what happens to the current in different circuits.

1 One way of explaining electrical circuits is to use a model. One model is the water model.

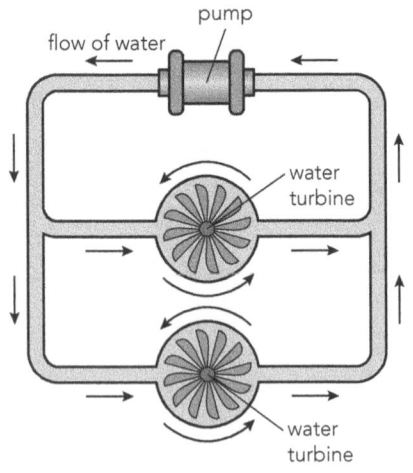

Water model for electrical circuits

Look at the diagram above. Explain the model. In what way does this model help us to understand series and parallel circuits? You may refer to the diagram in your answer.

_____

_____

_____

_____

2 This question is about the circuit shown in the diagram below.

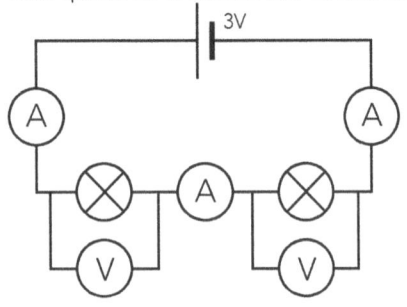

a One of the ammeters reads 0.50 A. What is the size of the current at the other ammeters?

_____

b What can you say about the size of the potential difference measured by each of the voltmeters?

_____

c What is the size of the potential difference across each lamp?

_____

3 This question is about the circuit in the diagram below.

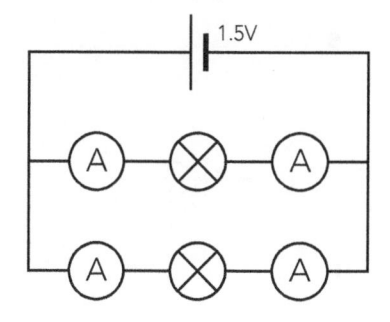

a One of the ammeters reads 0.50 A. What is the size of the current at the other ammeters?

_____

b What is the total current in the circuit?

_____

c There are no voltmeters in this circuit. Add one voltmeter to the diagram to measure the potential difference across either lamp.

d What should the reading on the voltmeter be?

_____

4 Describe the way in which the current and potential difference vary in different parts of a series circuit.

_____

_____

5 Describe the way in which current and potential difference vary in different parts of a parallel circuit.

_____

_____

6 Identify which component each of the following circuit symbols represents.

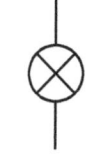

a _____   **b** _____   **c** _____

1 Electricity

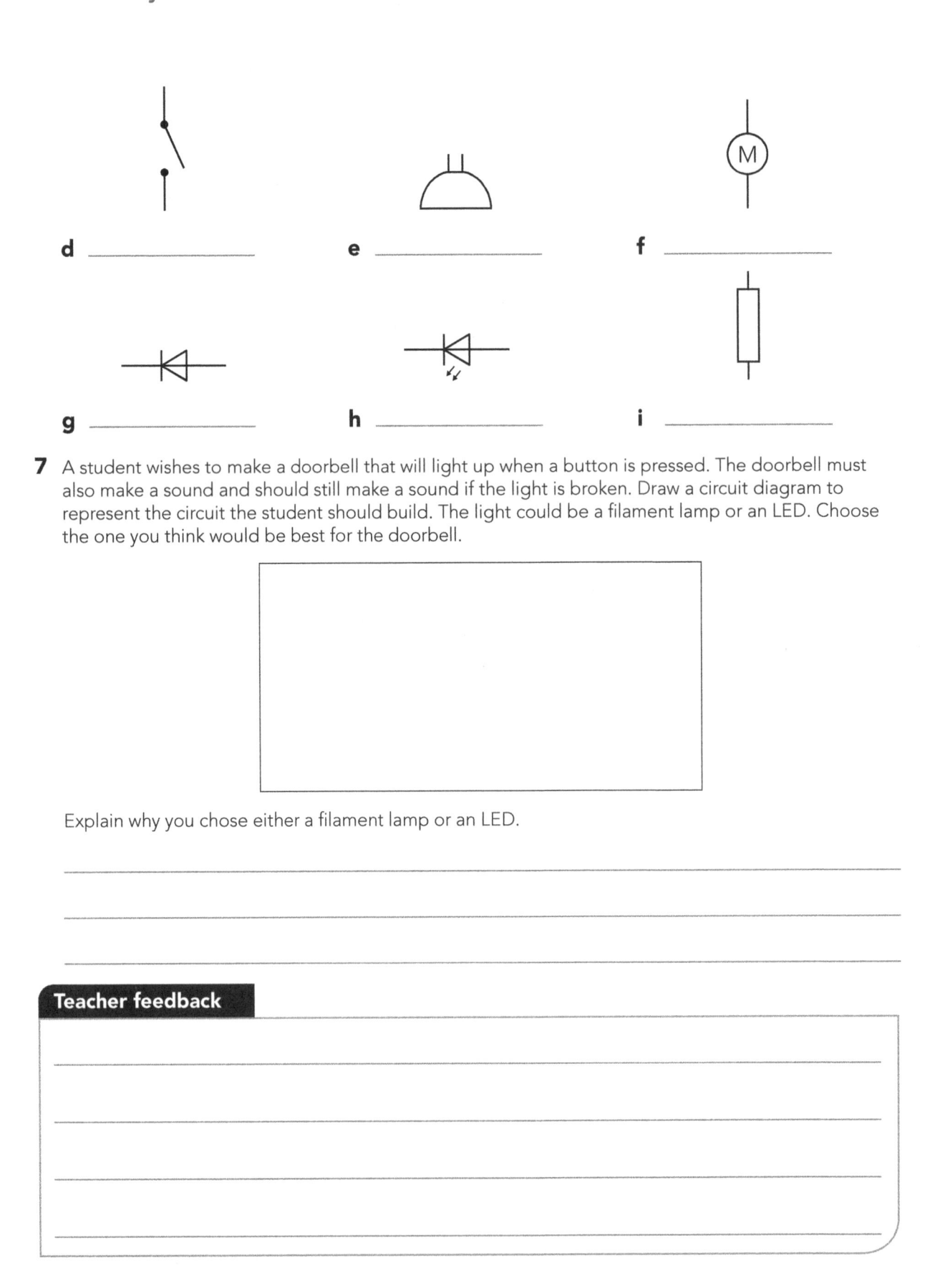

d _____   e _____   f _____

g _____   h _____   i _____

7 A student wishes to make a doorbell that will light up when a button is pressed. The doorbell must also make a sound and should still make a sound if the light is broken. Draw a circuit diagram to represent the circuit the student should build. The light could be a filament lamp or an LED. Choose the one you think would be best for the doorbell.

Explain why you chose either a filament lamp or an LED.

_____

_____

_____

Teacher feedback

Mains electricity
Student Book Topic 1.11

This exercise will require you to demonstrate an understanding of mains electricity and how it differs from the electricity that comes from a battery.

1 The name for the type of electricity that comes from a battery is often abbreviated to DC. State what DC stands for and explain what it means.

_____

_____

2 The name for the type of electricity that comes from the mains is often abbreviated to AC. State what AC stands for and explain what it means.

_____

_____

3 The following graph shows the difference between alternating current and direct current. Label the line showing alternating current and the line showing direct current.

4 Draw the circuit symbol for a fuse in the space below.

5 Fuses are a safety feature that can prevent electrical devices from catching fire. Explain the way in which a fuse protects electrical devices.

_____

_____

6 Depending on where you live, mains electricity can have a potential difference of up to 230 V. Use this to explain why mains electricity is dangerous, and explain what can happen if you are electrocuted.

_____

_____

_____

Teacher feedback

Chapter 2 Moments, pressure and density

Exercise 1 — Forces and simple machines
Student Book Topic 2.1

This exercise will allow you to see how forces can be used to transfer energy, and how we can use pulleys to increase the size of a force.

1. Identify which of the following situations involve work being done. Put a tick next to the situations where work is being done and a cross next to the situations where work is not being done.
   - A shelf holding up a stack of books.
   - A student throwing a ball.
   - A foot pump being pushed down.
   - A strong man leaning against a brick wall.
   - A weightlifter holding 40 kg above his head.
   - A railway porter carrying a passenger's two cases.
   - A railway porter holding the two cases waiting for a tip.

2. The weightlifter in question 1 has to lift the 40 kg mass from the ground to above his head. Gravity pulls down on a 40 kg mass with a force of 400 N. Calculate the amount of work the weightlifter has to do. Include a unit with your answer.

3. State how much gravitational potential energy the 40 kg mass has gained by being lifted over the weightlifter's head.

10

2 Moments, pressure and density

3 Pulleys are devices that can magnify a force. They do not increase the amount of work done though, so if the size of the force increases, the distance moved must decrease.

Here are four different arrangements of pulleys.

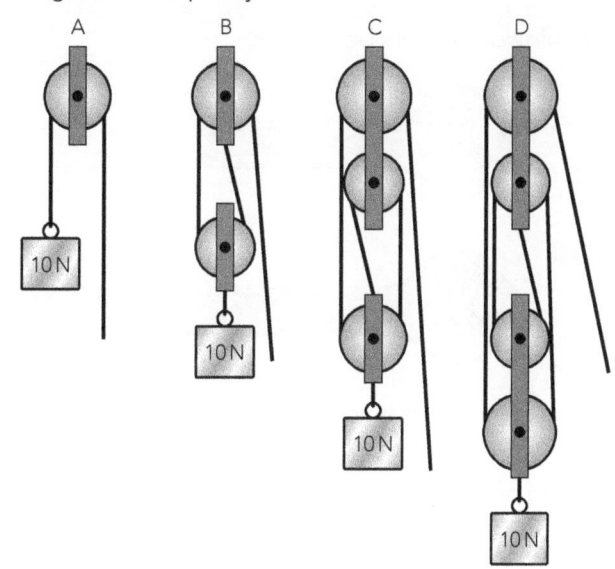

For each of the pulley systems, a student connects a newtonmeter to the end of the string and records the force needed to raise the 10 N weight. The student also records the distance the weight moves when the newtonmeter moves 30 cm.

Complete the table for each of the pulley systems. The first one has been done for you.

Pulley	Weight / N	Distance moved by weight / cm	Force required / N	Distance moved by newtonmeter / cm
A	10	30	10	30
B	10			30
C	10			30
D	10			30

Teacher feedback

Exercise 2

Levers and moments
Student Book Topics 2.2 and 2.3

This exercise is about levers and moments. You will gain an understanding of the factors that affect how well levers operate. You will also look at how spanners are designed to make them as effective as possible.

1 Have a look at the picture of two spanners below.

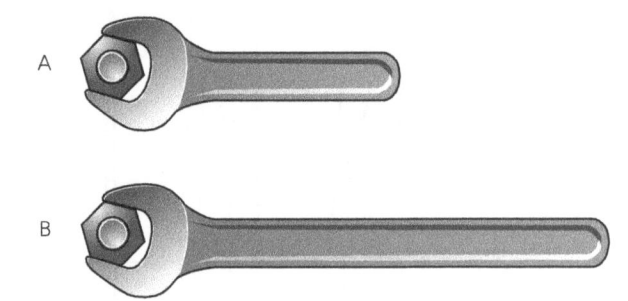

A

B

 a On each of the spanners draw an arrow to show where the force should be applied to loosen the nut, and in which direction.

 b Using your understanding of levers, explain why it would be easier to loosen the nut with the longer spanner, B.

 _____

 _____

2 Lots of objects and devices we use take advantage of levers. Levers make jobs easier. List as many objects as you can that are made from levers.

 _____

 _____

 _____

3 Complete the following sentences using the words in the box.

| moment | greater | pivot | force | smaller | force |

 The further a _____ acts from the pivot, the _____ the moment of the force.

 The smaller the _____, the _____ the moment of the force.

 The formula used to calculate moments is: _____ = force × distance from _____ .

2 Moments, pressure and density

4 Complete the table to show the moment of a force of 100 N when it acts 1 m, 2 m, 3 m, 4 m and 5 m from a pivot. You will also have to add the units to the table.

Distance from pivot / m	Force / _____	Distance / _____	Moment / _____
1			
2			
3			
4			
5			

5 For the following moments, calculate the distance from the pivot.

  a moment 20 Nm, force = 4 N

  _____
  _____

  b moment 150 Nm, force = 30 N

  _____
  _____

  c moment 70 000 Nm, force = 7 kN

  _____
  _____

Teacher feedback

Exercise 3: Moments and equilibrium

Student Book Topic 2.4

This exercise will help you understand what happens when moments are balanced. You will also look at how we can investigate moments, practising your investigative skills.

1 Two children are on a seesaw.

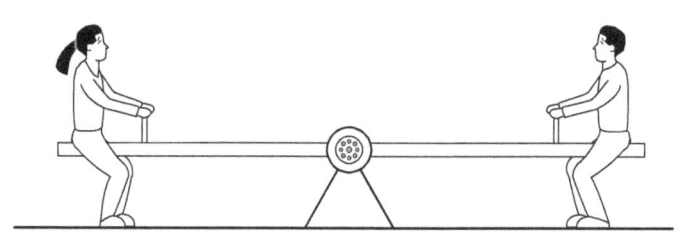

a On the diagram, add arrows to show the forces acting on the seesaw.

b On the diagram, label each of the forces.

c In which direction is the force due to the boy trying to turn the seesaw?

_____

d In which direction is the force due to the girl trying to turn the seesaw?

_____

e If the seesaw is not moving, what can you say about the turning effects of the forces?

_____

2 If all the moments on an object are balanced, an object will not turn. Write the principle of moments, using this information.

_____

_____

3 A student decides to investigate the principle of moments. The student does a preliminary (trial) experiment.

a Suggest some reasons why it is a good idea to do a preliminary experiment.

_____

_____

b During the preliminary experiment, the student balanced a meter rule on a triangular prism made of wood. The student then placed different weights at different distances from the pivot. Here are four of the arrangements the student used. For each one, predict whether or not they would balance. Circle the correct answers.

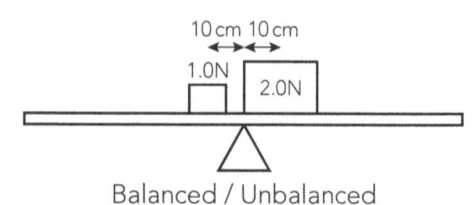

Left diagram: 1.0N at 10cm, 2.0N at 10cm — Balanced / Unbalanced

Right diagram: 1.5N at 30cm, 3.0N at 15cm — Balanced / Unbalanced

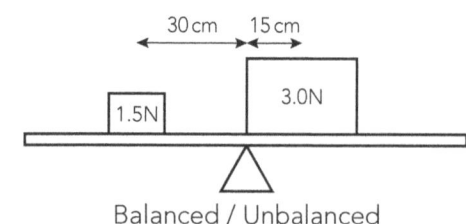

2 Moments, pressure and density

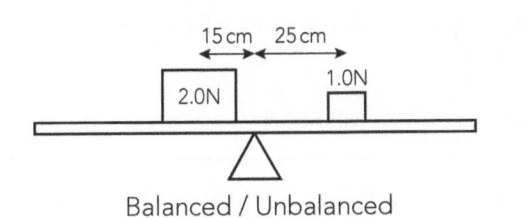

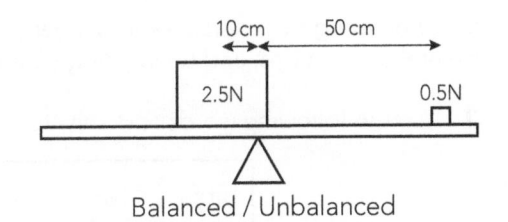

Balanced / Unbalanced   Balanced / Unbalanced

c After completing the preliminary experiment, the student draws the following table. This is to predict what should happen in the experiment. Fill in all the missing gaps to predict what should happen in the experiment when the ruler is balanced. The first row has been completed for you.

Anticlockwise (left hand side)			Clockwise (right hand side)		
Force / N	Distance / m	Moment / Nm	Force / N	Distance / m	Moment / Nm
1.0	0.10	0.10	2.0	0.05	0.10
1.0	0.20	0.20	2.0		
2.0	0.10	0.30		0.30	
2.0	0.20	0.40		0.10	
3.0	0.25	0.50	2.5		
3.0	0.30	0.60		0.40	

d When the experiment has been completed the student will draw a graph. **Sketch** a graph on the axes below to predict what the student's graph will look like. Do not plot any points on the graph, or add a scale.

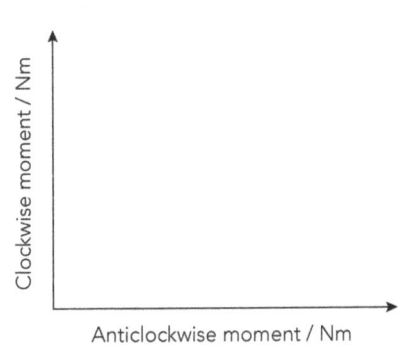

e When experiments are conducted, the data that is gathered rarely fits a neat pattern (and may contain some anomalous results). The data in the table and the graph you have drawn does fit a neat pattern. Suggest some reasons why any data gathered in an experiment will sometimes contain anomalies.

Teacher feedback

15

Exercise 4: Pressure on surfaces

Student Book Topic 2.5

By completing this exercise you will gain a better understanding of pressure. You will be performing several calculations, and developing your problem-solving skills.

1 Use the following words and units to complete the sentences. You will need to use one of them twice.

| area | large | force | Pa | N/m² | smaller | pressure |

Pressure tells you how concentrated a _____ is. Pressure is measured in _____ or _____ and is calculated using the formula: _____ (N/m²) = _____ (N) / _____ (m²).

If a force is acting over a small area, the pressure is _____ .

If the force acts over a larger area, the force is more spread out and so the pressure is _____ .

2 A car weighs 2 500 N. The area of the four wheels that is in contact with the road is 0.50 m². What is the pressure on the road?

_____

_____

3 A pile of bricks weighs 12 000 N which acts over an area of 2.0 m². What is the pressure?

_____

_____

4 A student weighs 450 N. The student's feet cover an area of 0.11 m². The student stands on ice that will crack if the pressure is greater than 6 500 Pa. Is the student safe on the ice? Show your calculations.

_____

_____

5 A car gets stuck in some mud. These are some things that the driver could do to try to get the car out of the mud.

A. Tell the passengers to get out of the car.

B. Keep pressing the accelerator to make the wheels spin faster.

C. Let some air out of the tyres.

D. Push the car.

Answer the following questions by writing the letter of the correct statement, A–D, in the spaces provided on the right. There may be more than one letter for each.

a Which of these would increase the area of the tyres? _____

b Which of these would make the pressure less? _____

c Which would reduce the force on the mud? _____

d Which would have no effect on the pressure? _____

e Which might make the car sink deeper into the mud? _____

Exercise 5: Pressure in liquids and gases

Student Book Topics 2.6 and 2.7

This exercise is about the pressure exerted by liquids and gases.

1 Read the following statements and decide if they are true or false.

Example	True or False?
Hydroelectric dams are thicker at the top to stop the dam from bursting.	
Divers need to wear specialist watches when diving in deep water.	
Water pressure decreases as the depth of the water increases.	
Pressure in liquids acts equally in all directions.	

2 Only one of the following diagrams is correct. Which one?

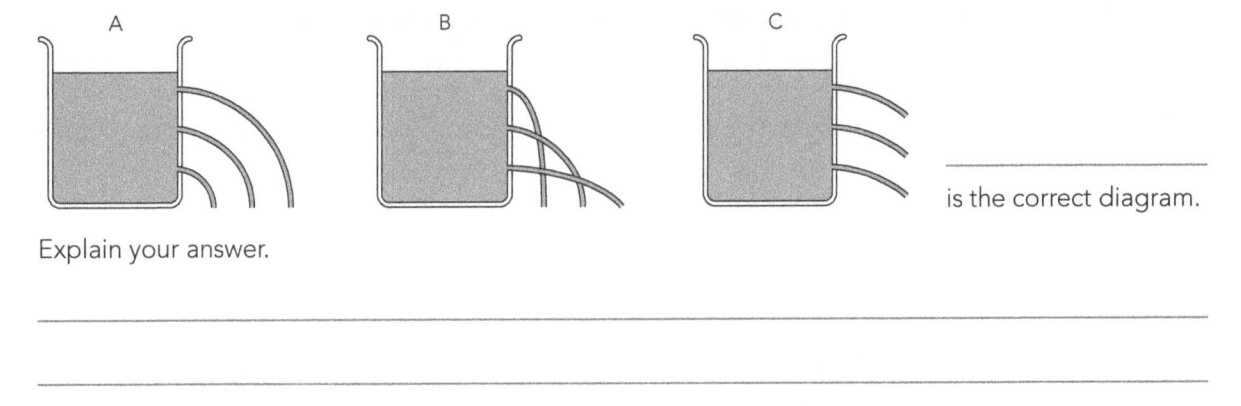

A   B   C

_____ is the correct diagram.

Explain your answer.

_____

_____

_____

3 Here is a deep-sea diver.

  a Add arrows to the diagram to show the direction that the pressure will act when the deep-sea diver is under water.

  b Explain what happens to the pressure on the diver as he descends deeper into the ocean.

_____

_____

_____

  c When returning to the surface, the diver is in danger of developing 'the bends'. Explain what the diver must do to prevent this from happening.

_____

_____

_____

_____

Exercise 6: Density of fluids and hydraulics

Student Book Topics 2.8 and 2.9

Before you start this exercise you may find it helpful to look again at the work you did at Stage 7 on densities of solids (Student book topic 10.8, Workbook Chapter 10, Exercise 4).
This exercise is designed to help you understand how we can measure the density of fluids. It will also help you understand how hydraulic systems work.

1 A student wishes to calculate the density of olive oil. The student measures out 500 cm³ of oil.

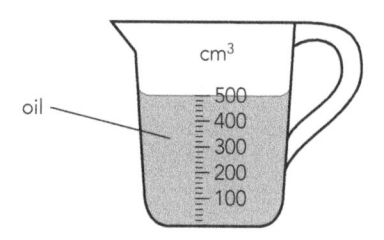

a What else would the student have to measure in order to calculate the density?

b Show the way in which the student would use both measurements to calculate the density of the oil.

c Fresh water has a density of 1 g/cm³. What is the mass of 500 cm³ of water? Give your answer in both grams and kilograms.

d Salt water has a higher density than fresh water. Salt water is usually about three percent denser than fresh water. Estimate the mass of 500 cm³ of salt water. Give your answer in both grams and kilograms.

2 The student now wishes to compare the density of different liquids. The student pours 300 g of each liquid into three different measuring cylinders (not drawn to scale).

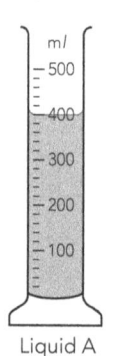

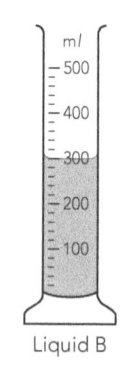

Liquid A Liquid B Liquid C

2 Moments, pressure and density

a Without performing any calculations, state which liquid has the highest density.
_____

b Which of the liquids is likely to be water?
_____

c A common type of cooking oil has a density of 0.8 g/cm³. Which of the liquids is cooking oil?
_____

3 Cooking oil and water do not mix. When put into a beaker, they form separate layers. Label the diagram opposite to show which layer is cooking oil and which layer is water.

4 Hydraulic systems work because any pressure applied to a liquid is transferred through the liquid. Complete the following paragraph, using the words in the box, to explain the way in which hydraulic systems work.

| small | larger | forces | pressure | larger |

Liquids can be used to transfer _____ to where they are needed. A force is applied to a small syringe. The force acts on a _____ area and puts the liquid under _____. The liquid then 'transfers' the same pressure to a bigger syringe with a _____ surface area. As the surface area is larger, the force produced will be _____.

5 a A small syringe has an area of 5 cm², and a larger syringe has an area of 25 cm². How many times bigger is the force on the larger syringe? Show your calculations.
_____
_____

b If the force on the small syringe is 20 N, what is the size of the force on the bigger syringe? Show your calculations.
_____
_____

Teacher feedback

19

Chapter 3 Thermal energy transfers and the energy crisis

Exercise 1 — Thermal energy
Student Book Topic 3.1

This exercise will help you understand the difference between thermal energy and temperature. You will also be considering different uses for thermometers and practising how to read them.

1 Thermometers are used to measure temperature. Record the temperature of each thermometer in the spaces provided.

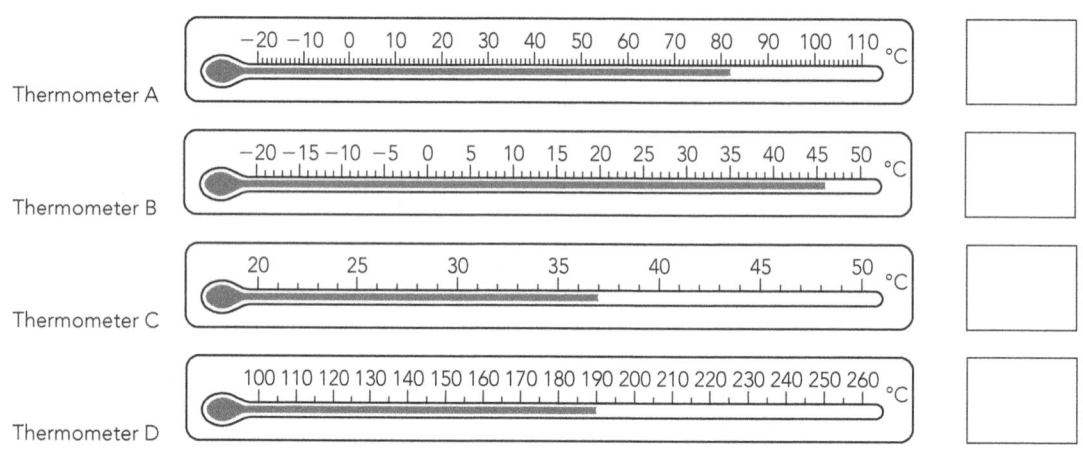

Thermometer A

Thermometer B

Thermometer C

Thermometer D

2 Different thermometers have different uses. Write the letter of the thermometer from question 1, next to a possible use for the thermometer.

- In an oven. _____
- In water during a heating experiment. _____
- Taking a person's temperature. _____
- As a weather thermometer. _____

3 Rank the following substances to show which ones contain the most energy. Number 1 will have the most energy and number 5 will have the least.

Substance	Mass / g	Temperature / °C	Rank
Tap water	35	15	
Ice	35	0	
Water from a kettle	35	100	
Air	35	25	
A mixture of water and ice	35	0	

4 A student heats up different volumes of water using the same Bunsen burner for 60 seconds. Explain why the different volumes of water will not have the same temperature rise.

_____

_____

_____

20

Exercise 2: Conduction and convection

Student Book Topics 3.2 and 3.3

This exercise will help you understand the processes of conduction and convection. By answering the questions you will be considering how heat is transferred through solids, liquids and gases.

1 Complete the following paragraph using the words in the box.

> other   good   kinetic   electrons   vibrate   close   solids

Conduction is the main form of heat transfer in _____. This is because the particles are very _____ together. Extra heat energy makes the particles _____ more. They pass on the extra vibrational energy to _____ particles. Another term for vibrational energy is _____ energy.

Metals are _____ conductors of heat energy because they contain many free _____.

2 Which materials are good conductors and which are good insulators? Complete the table with an example of each material and put a tick (yes) or a cross (no) in each box in the last two columns.

Type of material	Example	Good conductor?	Good insulator?
Solid metal			
Solid non-metal			
Liquid			
Gas			

3 Complete the following sentence by crossing out the incorrect words.

Heat energy travels through metals from places where the temperature is **high / low** to places where the temperature is **high / low**.

4 The following picture shows a metal rod being heated in a Bunsen burner.

Explain the process in which conduction transfers heat from one end of the metal rod to the other.

_____

_____

_____

_____

_____

21

3 Thermal energy transfers and the energy crisis

5 The diagram below shows what happens when you heat a pan of water. Use the space provided to describe the way in which heat is transferred through the water. Your description should contain the key words in the box.

> dense   sinks   convection
> heat   rises   expands   particles

Convection

_____
_____
_____
_____

6 Use your understanding of convection to answer the following questions.

   a Why is an air-conditioning unit placed close to the ceiling?

   _____
   _____

   b Why is a heater located close to the floor?

   _____
   _____

   c Why should you crawl near to the floor in a smoke-filled room?

   _____
   _____

7 There are three states of matter. In which state is it **not** possible for heat to be transferred by convection.

_____

8 Space is a vacuum. Explain why heat cannot be transferred from the Sun to the Earth by conduction or convection.

_____

Exercise 3

Radiation

Student Book Topic 3.4

This exercise is about how heat can be transferred by infra-red radiation. You will also need to consider conduction and convection to answer some of the questions.

1. Different surfaces emit, absorb and reflect infra-red radiation differently. Write the word '**good**' or '**bad**' in each of the spaces in the table.

Type of surface	Emitter	Absorber	Reflector
Dark and dull			
Light and shiny			

2. Although infra-red radiation is very similar to light, the human eye cannot detect infra-red radiation. Some animals can detect infra-red radiation. Name an animal that can, and suggest why this is an advantage to the animal.

   _____

   _____

3. In hot countries houses are often painted white. Explain the effect that this has on the temperature inside the house:

   a during the day.

   _____

   _____

   b during the night.

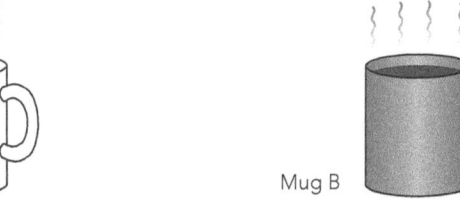

   _____

   _____

4. Look at the two mugs below. Which mug would keep tea hot for the longest time? Explain your answer.

   Mug A   Mug B

   _____

   _____

   _____

5. State two uses of infra-red radiation.

   _____

3 Thermal energy transfers and the energy crisis

6 Next to each statement, write one of the terms in the box below to explain if the statement refers to conduction, convection, radiation or all three.

conduction	convection	radiation	all three

a Heat flowing between two places when there is a difference in temperature. _____

b Energy passing from atom to atom. _____

c Heat transfer occurring through transparent substances. _____

d Sets up movement currents in liquids and gases. _____

e Is affected by colour and how shiny a surface is. _____

f Can occur through a vacuum. _____

g Involves hot fluid expanding and rising. _____

7 Some of the following statements are true and some are false. Write **true** or **false** next to each statement.

a Heat energy is transferred through solids by conduction. _____

b The air is a good conductor of heat. _____

c A poor conductor is called an insulator. _____

d All solid objects are good conductors of heat. _____

e Convection currents occur because particles expand. _____

f Convection is a way of transferring heat energy by the movement of particles. _____

g Liquids and gases become less dense when they are heated. _____

h Convection can sometimes occur in solid objects that are hot enough. _____

i Not all objects emit infra-red radiation. _____

j A cold cup of tea will emit infra-red radiation more quickly than a hot cup of tea. _____

k A dark, matt surface will emit infra-red radiation at a faster rate than a light, shiny surface at the same temperature. _____

l Hotter objects emit infra-red radiation at the same rate as similar cold objects. _____

Teacher feedback

Exercise 4: Evaporation

Student Book Topic 3.5

You will be able to demonstrate your understanding of evaporation in this exercise, and the effect evaporation has on temperature.

1. Evaporation occurs when particles that are moving fast enough escape from the surface of a liquid. This can occur at any temperature. List the four factors that can increase the rate of evaporation.

   1 _____   3 _____

   2 _____   4 _____

2. Wet clothes can be hung on a washing line to dry them. The weather conditions affect how long it takes them to dry. Answer the following questions about drying clothes on a washing line.

   a Why are warm days the best days to hang out your washing?

   _____

   _____

   b Why are windy days the best days to hang out your washing?

   _____

   _____

   c Why should washing be stretched out on the washing line?

   _____

   _____

3. When the most energetic particles leave the surface of the liquid, what effect does this have on the remaining liquid?

   _____

4. Use what you know about evaporation to explain why people sweat when they are hot.

   _____

   _____

   _____

5. Some animals cannot sweat. Other than bathing in water, what else can such an animal do to cool down on a hot day?

   _____

   _____

   _____

Exercise 5: Fuels for electricity

Student Book Topic 3.6

In this exercise you will be reinforcing your learning about where different fuels come from and looking at alternative fuels.

1 Name the **three** different types of fossil fuels.

_____

2 The following sentences explain one way in which fossil fuels are formed. Reorder the sentences (1–5) in the spaces provided.

- The weight of the mud pushed down on the dead animals. _____
- The dead animals turned into fossil fuels over millions of years. _____
- Millions of years ago, animals died and sank to the sea-bed. _____
- We dig up the fossil fuels and use them as a source of energy. _____
- Layers of mud built up on top of the dead animals. _____

3 What are biofuels?

_____

4 Biofuels can be described as being carbon neutral. Explain what the phrase 'carbon neutral' means.

_____

_____

5 A student wishes to investigate how much energy various fuels release when they are burned. He has been given an equal mass of different fuels in spirit burners.

He decided to test the fuels by seeing which fuel causes the biggest temperature rise in a certain volume of water.

a The first thing he does is to carry out a risk assessment. What are the risks associated with the investigation? Include what the student should do to minimise the risk.

Spirit burner

_____

_____

b Write a list of the equipment he would need.

_____

_____

_____

3 Thermal energy transfers and the energy crisis

c Draw a labelled diagram to show the way his experiment was set up.

d Here are the student's results from the experiment. Complete the final column.

Name of fuel	Temperature before / °C	Temperature after / °C	Temperature rise / °C
Ethanol	20	46	
Propane	20	52	
Vegetable oil	21	42	
Diesel	20	61	

e Should the student draw a bar chart or a line graph for this data?

f What should the title of the bar chart or line graph displaying his results be?

g Draw a bar chart or line graph for the data.

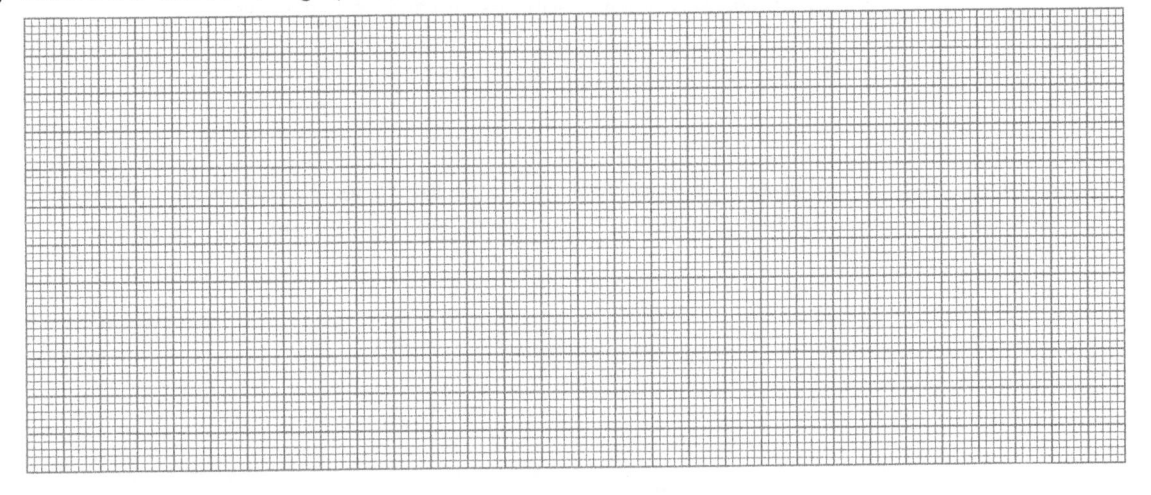

6 The student burned each fuel for the same length of time. He noticed that with different fuels, the flame was not the same size. The student realised that his experiment was not a fair test. Suggest what needs to be done to make the experiment a fair test.

Exercise 6: Renewable energy resources

Student Book Topics 3.7 to 3.10

This exercise will help you identify different renewable energy resources.

1 The following table contains a list of statements. Each one refers to a renewable energy resource. Write the names of the energy resources in the second column.

Statement	Energy resource
Destroys the habitat of wading birds	
Involves damming upland river valleys	
Hot water and steam from volcanic rocks are used to drive turbines	
Causes visual and noise pollution and depends on air currents	
Floods land that could be used for farming or forestry	
A renewable energy source that is very reliable and involves fast-flowing water	
Involves building a tidal barrage across a river estuary	
The amount of electricity produced depends on the state of the tides	
The amount of electricity produced depends on light intensity	
Energy released from the decay of radioactive substances within the Earth	
Can be operated in reverse to pump water back to high levels using surplus electricity during the night	

2 Hydroelectric power plants use water that has fallen as rain. The rainwater is free. Explain why generating electricity using renewable energy resources can still be expensive.

3 Nuclear power stations use uranium as their fuel. Very little fuel is needed because it releases so much energy. This makes the fuel relatively cheap. Other than building the power station, explain why nuclear power stations are expensive.

Teacher feedback